ILLUSTRATED CATALOGUE OF SETH THOMAS

Clocks, Regulators

AND

TIME PIECES,

MANUFACTURED AND SOLD BY THE

Seth Thomas Clock Company

Plymouth Hollow, Conn.

New York:
FRANCIS & LOUTREL, PRINTERS, 45 MAIDEN LANE,

1863.

WITH ADDITIONS 1864 - 1872
ALSO
TOWER CLOCKS, 1874

THE 1863 CATALOG IS REPRODUCED IN ENTIRITY. THE 1864 CATALOG WAS IDENTICLE TO THIS CATALOG, EXCEPTING THE CHANGE OF DATE ON THE TITLE PAGE. THE SIZE OF THE ORIGINAL OF THESE CATALOGS WAS 6 1/4 X 4 INCHES.

INTRODUCTION

During February, 1973 we reproduced one thousand copies of a digest of the earliest catalogs issued by the Seth Thomas Clock Company. With the first printing depleted, we offer this second edition of fifteen hundred copies with little variation. The cover of this printing is a facsimile of the 1863 edition.

Dated 1863, the earliest known catalog for this firm is included in entirety and supplemented with the addition of new models from the 1866 and 1868 editions. Their earliest known tower clock catalog, 1874, is reproduced in entirety and selections from the 1872 catalog of Seth Thomas' Sons & Company, an allied firm, are also included.

A documented history of this important clockmaking firm has been added to the final pages of this publication. Research was initiated to determine the years each model was dropped from the line and this information is noted at the bottom of each page.

We gratefully acknowledge the loan of the 1866 catalog by the Connecticut Historical Society. All other original materials herein are from the Edward Ingraham Library of the American Clock & Watch Museum, Inc. We invite your support and membership in the Museum and participation in its programs.

July 1977

Chris H. Bailey,
Managing Director
American Clock & Watch Museum, Inc

ISBN 0-930476-00-X (FORMERLY 0-913602-07-8)
LIBRARY OF CONGRESS CATALOG CARD NUMBER 72-97555

PUBLISHED BY THE
AMERICAN CLOCK AND WATCH MUSEUM, INC.
100 MAPLE ST., BRISTOL, CONN. 06010

PRINTED BY THE BOND PRESS, INC., HARTFORD, CONN.

MANY of the Clocks represented on the following pages have been in the market so many years, and are so widely known, that a particular description is unnecessary; but having recently commenced the manufacture of several new styles, we have found it necessary, for the convenience of our customers, to issue this Catalogue.

That our business has throughout a half a century steadily increased from a small beginning to its present dimensions, is sufficient evidence of the quality of our work.

Our CALENDAR CLOCKS are reliable Time Keepers and Calendars combined, as several thousands of them, running correctly and satisfactorily in all parts of the world, sufficiently attest.

They require the winding of the time movement only to secure the required changes at midnight; showing the day of the week, the month, and day of the month, (including the 29th day of February of leap year,) perpetually, and with unfailing certainty.

We are sole owners of the patents under which we are manufacturing these Calendars, and are prepared to put them up to order, with months and days of week in French or Spanish.

No pains will be spared to maintain and increase the reputation of the "SETH THOMAS CLOCKS," and we give our warrant to the full amount of the cost, that every Clock that leaves our factory, if properly used, will prove a good Time Keeper.

SETH THOMAS CLOCK CO.

NOTE.—These Clocks are all drawn on a scale of 1½ inches to the foot, or one-eighth the size of the original.

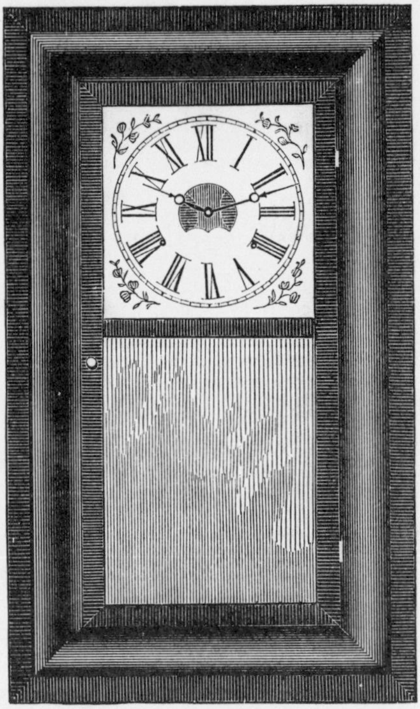

8 DAY.

O. G.—WEIGHT—STRIK

Height of Clock 29½ inches.

8 DAY.

Rosewood—Column—Strike—Weight.

Height of Clock 32 inches.

8 DAY.

Rosewood—Shell Column—Strike—Weight.

Height of Clock 32 inches.

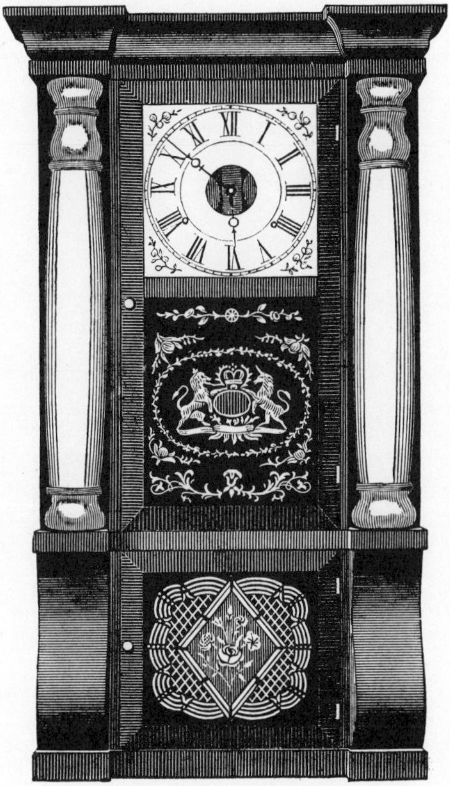

8 DAY.

Rosewood—Gilt or Gilt Cap—Strike—Weight.

Height of Clock 32 inches.

30 HOUR.

O. G.—STRIKE—WEIGHT.

Height of Clock 25 inches.

D. 1913
ALSO OFFERED AS A SMALL 8 DAY IN 1868. THIS SMALL 8 DAY WAS REPLACED BY AN 8 DAY OOG IN 1874.

30 HOUR.

Rosewood Column—Strike—Weight

Height of Clock 25 inches.

D. 1909

ALSO OFFERED AS A SMALL 8 DAY IN 1868 (D. 1880)

30 HOUR.

Rosewood Shell Column—Strike—Weight.

Height of Clock 25 inches.

D. 1909

ALSO OFFERED AS A SMALL 8 DAY IN 1868 (D. 1880)

30 HOUR.

Rosewood—Gilt or Gilt Cap—Column—Strike—Weight.

Height of Clock 25 inches.

D. 1909

ALSO OFFERED AS A SMALL 8 DAY IN 1868 (D. 1880)

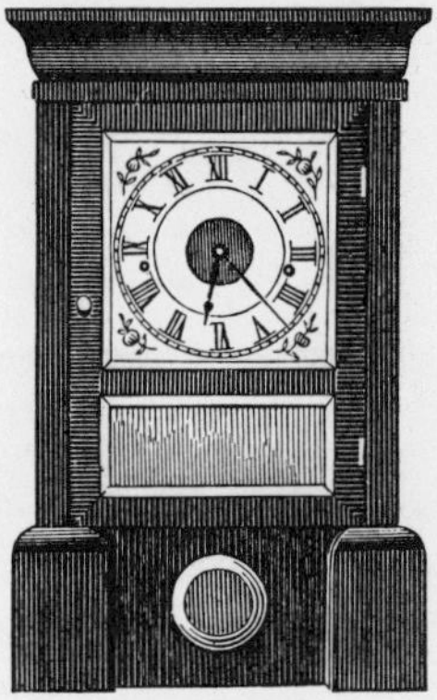

30 HOUR.

O. G. Top—Spring—Strike.

Height of Clock 15 inches.

D. 1874 OR 1876

ALSO OFFERED AS AN 8 DAY IN 1866

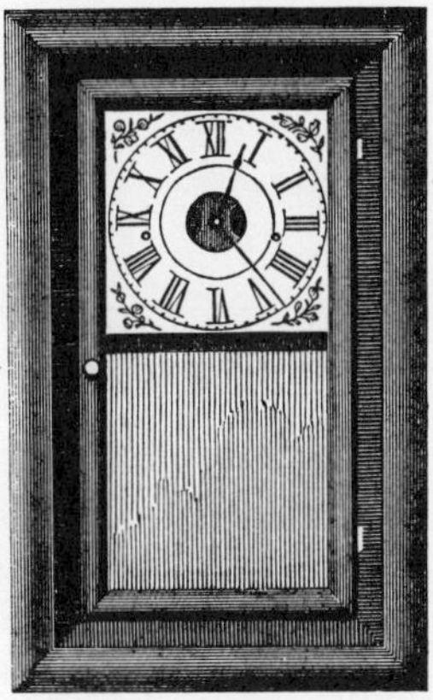

8 DAY OR 30 HOUR.

Round Band—Spring—Strike.

Height of Clock 17 inches.

8 DAY OR 30 HOUR.

Cottage—Spring—Strike.

Height of Clock 14 inches.

D. 1917

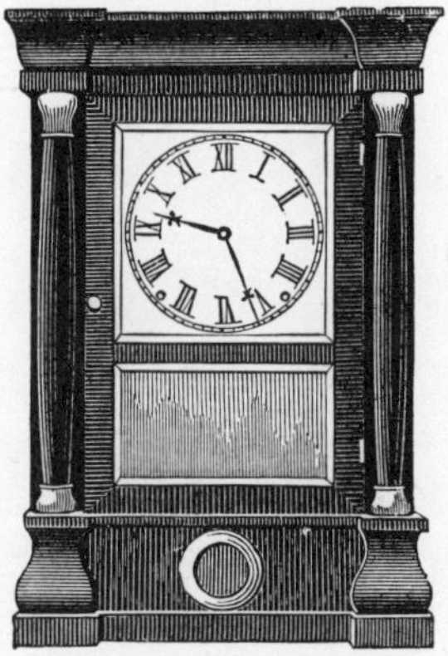

8 DAY.
COLUMN—SPRING—STRIKE.
Height of Clock 16 inches.

D. 1890

ALSO OFFERED AS A 30 HOUR IN 1868

8 DAY OR 30 HOUR.

PARLOR—SPRING—STRIKE.

Height of Clock 14 inches.

8 DAY or 30 HOUR.

OCTAGON TOP—SPRING—STRIKE.

Height of Clock 14 inches.

Cottage Time Piece.
WITH OR WITHOUT ALARM.
Height of Clock 7½ inches.

D. 1898 OR 1900
OFFERED AS 30 HOUR OR 8 DAY IN 1868
ALSO OFFERED AS 30 HOUR WITH STRIKE IN 1868

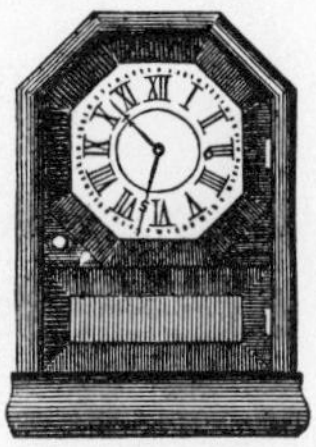

Octagon Time Piece.

WITH OR WITHOUT ALARM.

Height of Clock 7½ inches.

8 DAY.

PARLOR—STRIKE—WEIGHT.

Height of Clock 30½ inches.

No. 1.–8 DAY REGULATOR.

TIME—WEIGHT.

Height of Clock 34 inches.

D. 1890

TABLET DROPPED 1879

No. 2.–8 DAY REGULATOR.

TIME—WEIGHT.

Height of Clock 34 inches.

D. 1950

SECONDS BIT ADOPTED 1879. TABLET DROPPED 1880

8 DAY REGULATOR.

STRIKE—WEIGHT.

Height of Clock 44 inches.

D. 1879

CASE WAS MODIFIED AND INTRODUCED AS REGULATOR NO. 3 WITHOUT TABLET IN 1879. NO. 3 WAS D. 1917 THEN REVIVED IN 1928, BUT D. AGAIN IN 1930.

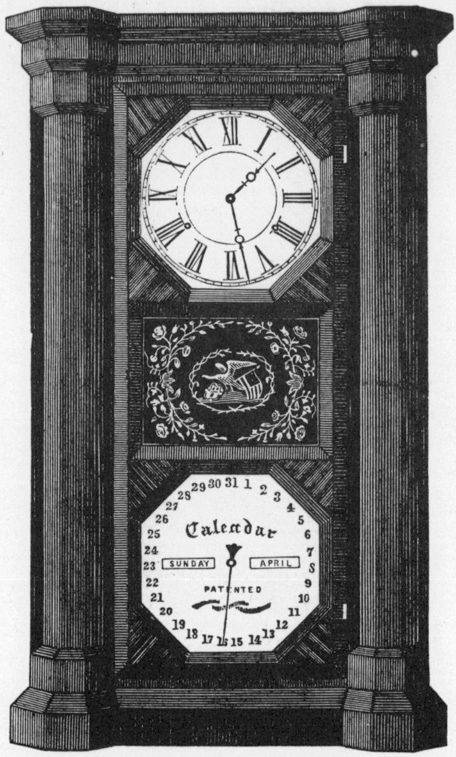

8 DAY PARLOR CALENDAR.

STRIKE—WEIGHT.

Height of Clock 30½ inches.

D. 1870 OR 1872

8 DAY.

OFFICE CALENDAR, No. 1—TIME—WEIGHT.

Height of Clock 40 inches.

8 DAY.

OFFICE CALENDAR, No. 2—TIME—WEIGHT.

Height of Clock 42½ inches.

Clock Trimmings.

Eight Day Thumb Keys, . .	per doz.
One Day " " . .	"
Eight Day, Crank " . .	"
One Day, " " . .	"
Pendulum Rods, ...	"
Hands, pairs, ...	"
Wire Bells, ...	"
Eight Day Springs, ...	"
One Day "	"
Verges, ...	"
Clock Cord, ...	per lb.
Bushings, Washers, &c., . .	"

ILLUSTRATED CATALOGUE

OF

SETH THOMAS

Clocks, Regulators,

AND

TIME PIECES,

MANUFACTURED BY THE

SETH THOMAS CLOCK CO.,

THOMASTON, CONN.

AMERICAN CLOCK COMPANY,

E. C. HINE, S. E. THOMAS,

Sole Agents,

No. 3 CORTLANDT STREET, NEW YORK,

AND

115 LAKE ST., CHICAGO.

THE 1866 CATALOG CONTAINED ALL MODELS PREVIOUSLY SHOWN IN THE 1863 AND 1864 CATALOGS EXCEPTING THE PARLOR MODEL, PAGE 16, WHICH WAS DISCONTINUED. THE FOLLOWING PAGES SHOW MODELS THAT WERE ADDED IN 1866. THE SIZE OF THE ORIGINAL OF THIS CATALOG WAS 6 1/4 X 4 INCHES.

8 DAY PARLOR—No. 1.

STRIKE—WEIGHT.

Hight, 32 inches.

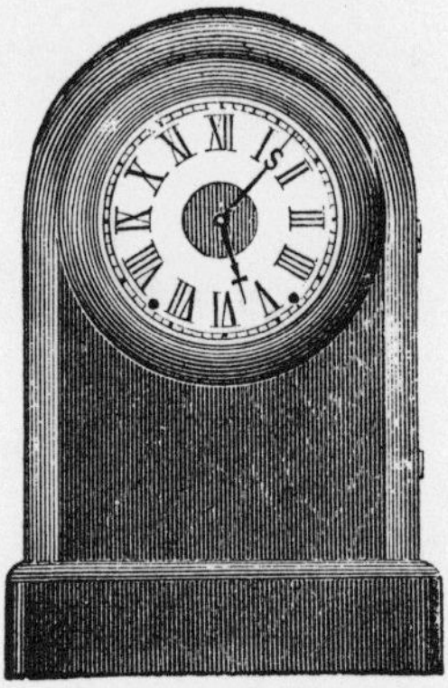

8 DAY.

TUDOR, No. 1.—Spring—Strike.

Hight, 16 inches.

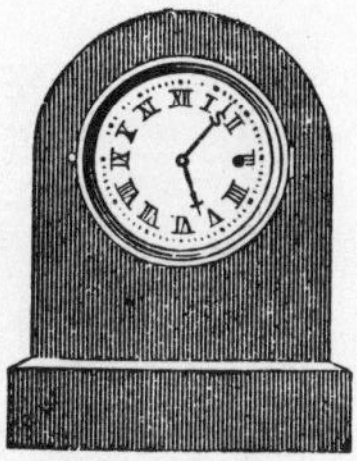

8 DAY CABINET TIME–No. 1.

(With or without Alarm.)

THIRTY HOUR—CABINET STRIKE—No. 1.

Height of Clock 9½ inches.

8 DAY CABINET TIME –No. 2.

(With or without Alarm.)

THIRTY HOUR—CABINET STRIKE—NO. 2.

Height of Clock 9½ inches.

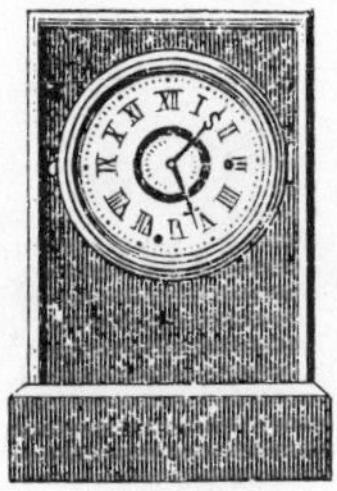

8 DAY CABINET TIME.–No. 3.
(With or without Alarm.)
THIRTY HOUR—CABINET STRIKE—No. 3.
Height of Clock 10 inches.

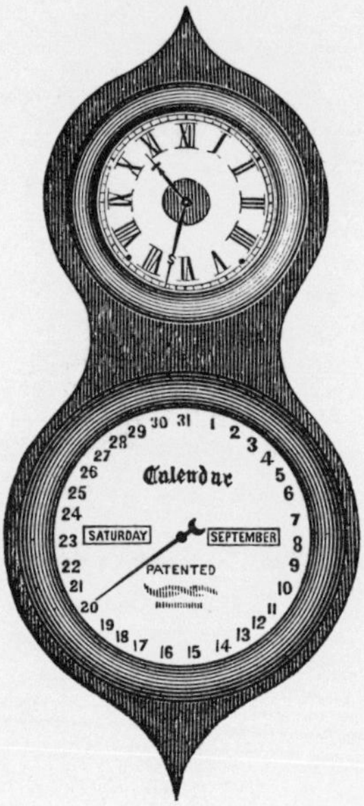

OFFICE CALENDAR.—No. 3.

8 Day—Spring—Time—6 inch Time Dial.

Hight, 24 inches.

D. 1870 OR 1872

8 DAY—PARLOR CALENDAR—No. 1.

STRIKE—WEIGHT.

Hight, 30½ inches.

Illustrated Catalogue

OF

SETH THOMAS

CLOCKS, REGULATORS

AND

TIME PIECES,

MANUFACTURED BY THE

SETH THOMAS CLOCK COMPANY,

THOMASTON, CONN.

Sole Agents,

AMERICAN CLOCK COMPANY,

3 Cortlandt Street, New York,

115 Lake Street, Chicago,

310 Sansome Street, San Francisco.

NEW YORK:
J. W. ORR & CO., ENGRAVERS AND PRINTERS, 96 NASSAU STREET.
1868.

THE 1868 CATALOG CONTAINED ALL MODELS SHOWN IN THE 1866 CATALOG EXCEPTING THE THREE CABINET MODELS SHOWN ON PAGES 31, 32, AND 33 WHICH WERE DISCONTINUED. THE FOLLOWING PAGES SHOW MODELS THAT WERE ADDED IN 1868. THE SIZE OF THE ORIGINAL OF THIS CATALOG WAS 6 3/4 X 4 1/4 INCHES.

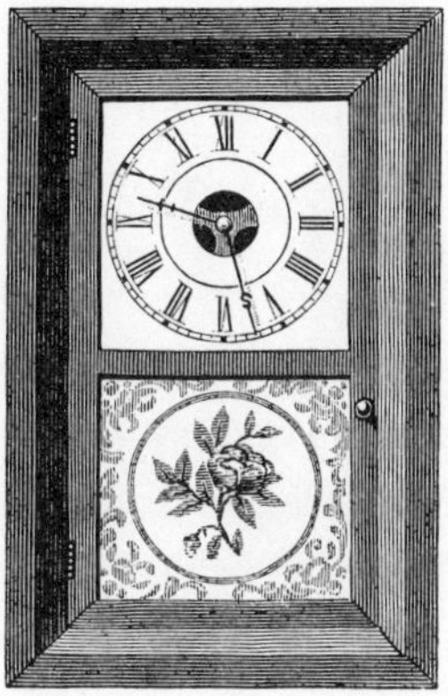

8 Day or 30 Hour.

KITCHEN—Spring—Strike.

Hight, 16 inches.

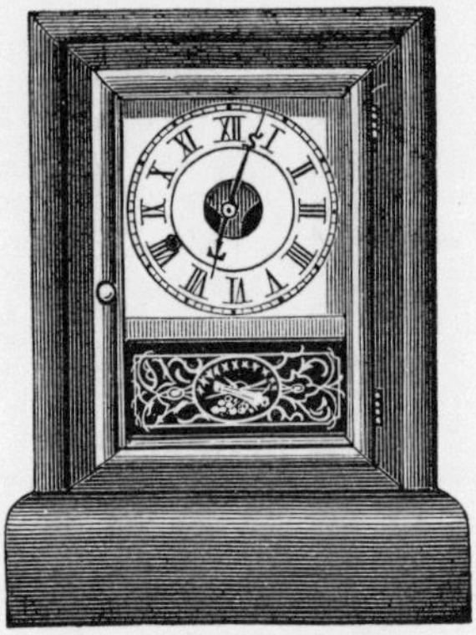

8 Day or 30 Hour.

COTTAGE, Extra—Spring—Strike.

Hight, 15 inches.

8 DAY OR 30 HOUR.

GOTHIC—Spring—Strike.

Hight, 15¾ inches.

(8 Day Strikes Half-Hours.)

8 Day.

TUDOR, No. 3.—Spring—Strike.

(Strikes Half-Hours.)

Hight, 12 inches.

8 DAY CABINET. Time—No. 1.—Extra.

30 HOUR CABINET. Strike—No. 1.—Extra.

Hight, 9½ inches.

8 DAY OFFICE.

SPRING—TIME OR STRIKE. 12 INCH DIAL.

Hight, 25 inches.

D. 1890

LATER CALLED OFFICE NO. 1

OFFICE CALENDAR.—No. 4.

8 DAY—SPRING—TIME—12 INCH DIAL.

Hight, 28 inches.

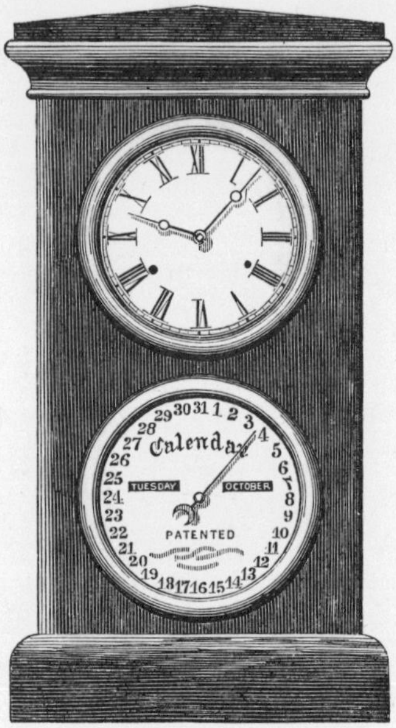

8 DAY PARLOR CALENDAR—No. 3.

STRIKE—SPRING.

Hight, 27 inches.

PRICE LIST
JANUARY 1, 1873

UNFORTUNATELY, 1873 IS THE EARLIEST PRICE LIST LOCATED FOR SETH THOMAS CLOCK CO. FROM THE JANUARY 1, 1873 LIST HAS BEEN RECONSTRUCTED, WHEN POSSIBLE, PRICES OF CLOCKS SHOWN IN THE PRECEEDING PAGES. PRICES WOULD HAVE BEEN SLIGHTLY DIFFERENT 1863–1868. LATER PRICE LISTS SHOW PRICES WERE DECREASING DURING THE YEARS 1873–1880.

PAGE		MODEL	PRICE
4	8 DAY	OG	$ 8.75
5	8 DAY	ROSEWOOD COLUMN	11.00
6	8 DAY	SHELL COLUMN	11.50
7	8 DAY	GILT COLUMN	12.00
8	30 HR	OG	6.00
8	8 DAY	OG (SMALL) (OFFERED 1868)	7.50
9	30 HR	ROSEWOOD COLUMN	7.00
9	8 DAY	ROSEWOOD COLUMN (SMALL) (OFFERED 1868)	8.50
10	30 HR	SHELL COLUMN	7.25
10	8 DAY	SHELL COLUMN (SMALL) (OFFERED 1868)	8.75
11	30 HR	GILT COLUMN	7.50
11	8 DAY	GILT COLUMN (SMALL) (OFFERED 1868)	9.00
12	30 HR	OG TOP	5.25
12	8 DAY	OG TOP (OFFERED 1866)	6.75
13	30 HR	ROUND BAND	4.50
13	8 DAY	ROUND BAND	6.00
14	30 HR	COTTAGE	4.25
14	8 DAY	COTTAGE	5.75
15	30 HR	COLUMN (OFFERED 1868)	6.50
15	8 DAY	COLUMN	8.00
16	30 HR	PARLOR	D. 1866
16	8 DAY	PARLOR	D. 1866
17	30 HR	OCTAGON TOP	5.50
17	8 DAY	OCTAGON TOP	7.00
18	30 HR	COTTAGE TIMEPIECE	2.75
18	8 DAY	COTTAGE TIMEPIECE (OFFERED 1868)	3.25
18	30 HR	SMALL COTTAGE STRIKE (OFFERED 1868)	3.75
19	30 HR	OCTAGON TOP TIMEPIECE	2.75
20	8 DAY	PARLOR	D. 1870–72
21	8 DAY	REGULATOR NO. 1	20.00
22	8 DAY	REGULATOR NO. 2	16.00
23	8 DAY	REGULATOR NO. 1 EXTRA	25.00
23	8 DAY	REGULATOR NO. 1 EXTRA, STRIKE	27.00
24	8 DAY	PARLOR CALENDAR (NO. 2)	D. 1870–72
25	8 DAY	OFFICE CALENDAR (NO. 1)	40.00
26	8 DAY	OFFICE CALENDAR NO. 2	35.00
MODELS ADDED IN 1866 –			
29	8 DAY	PARLOR NO. 1	18.00
30	8 DAY	TUDOR NO. 1	10.00
31	30 HR	CABINET NO. 1	D. 1868
31	8 DAY	CABINET NO. 1 TIMEPIECE (OR ALARM)	D. 1868
32	30 HR	CABINET NO. 2	D. 1868
32	8 DAY	CABINET NO. 2 TIMEPIECE (OR ALARM)	D. 1868
33	30 HR	CABINET NO. 3	D. 1868
33	8 DAY	CABINET NO. 3 TIMEPIECE (OR ALARM)	D. 1868
34	8 DAY	OFFICE CALENDAR NO. 3	D. 1870–72
35	8 DAY	PARLOR CALENDAR NO. 1	25.00
MODELS ADDED IN 1868 –			
37	30 HR	KITCHEN	4.25
37	8 DAY	KITCHEN	5.75
38	30 HR	COTTAGE EXTRA	4.25
38	8 DAY	COTTAGE EXTRA	6.25
39	30 HR	GOTHIC	5.50
39	8 DAY	GOTHIC	7.00
40	8 DAY	TUDOR NO. 3	8.00
41	30 HR	CABINET NO. 1 EXTRA	4.75
41	8 DAY	CABINET NO. 1 EXTRA TIMEPIECE	4.25
42	8 DAY	OFFICE NO. 1 TIMEPIECE	9.00
42	8 DAY	OFFICE NO. 1 STRIKE	10.50
43	8 DAY	OFFICE CALENDAR NO. 4	15.00
44	8 DAY	PARLOR CALENDAR NO. 3	15.00

SETH THOMAS' SONS & CO.'S

CATALOGUE

OF

FINE

MANTEL CLOCKS.

AMERICAN CLOCK COMPANY,

SOLE AGENTS,

No. 3 Cortlandt Street, New York.

No. 587 Wabash Avenue, Chicago. No. 520 Market Street, San Francisco.

SELECT PAGES FROM THE 1872 CATALOG ARE REPRODUCED. THESE INCLUDE PAGES 5, 11, AND 41 FROM THE ORIGINAL CATALOG OF 51 PAGES. THE SIZE OF THE ORIGINAL OF THIS CATALOG WAS 6 1/4 X 10 INCHES.

SETH THOMAS' SONS & CO.'S CATALOGUE.

We call attention to the Fine Mantel Clocks represented upon the following pages. We claim for them that they are fully equal to French Clocks in beauty of design and fineness of finish; that they are superior to the French Clocks as time-keepers, and are furnished at lower prices.

They require winding but twice monthly.

They strike the hours and half hours.

They have a noiseless tick, and a pleasant, clear-toned bell.

They are warranted good time-keepers.

We are constantly bringing out new designs, and are making a line of Bronze Figures and Ornaments to accompany these Clocks, of which we shall soon issue a Catalogue.

SETH THOMAS' SONS & CO.

New York, April 1st, 1872.

No. 5000. Hight, 10½ inches.

No. 5001.
Hight, 11 inches.

No. 5002.
Hight, 10 inches.

No. 8003. Hight, 18 inches.

COST JANUARY 1, 1879 WAS $18.00

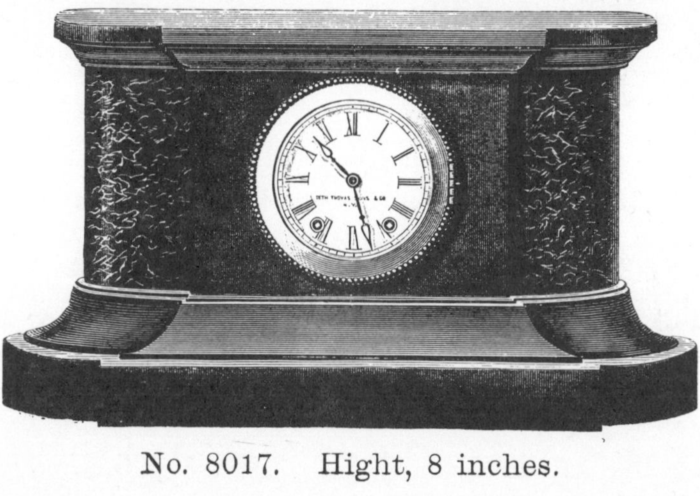

No. 8017. Hight, 8 inches.

No. 8064.
Hight, 19 inches.

No. 8066.
Hight, 15 inches.

COST JANUARY 1, 1879

$15.00 $19.50

CATALOGUE

OF

A. S. HOTCHKISS'

TOWER CLOCKS,

MANUFACTURED BY

SETH THOMAS CLOCK CO.

THOMASTON, CONN.

AMERICAN CLOCK CO.

AGENTS,

581 BROADWAY, NEW YORK.

172 State Street,
CHICAGO.

520 Market Street,
SAN FRANCISCO.

THE 1874 TOWER CLOCK CATALOG IS REPRODUCED IN ENTIRITY.
THE SIZE OF THE ORIGINAL OF THIS CATALOG WAS 9 1/2 X 6 IN.

THE TOWER CLOCKS

Designed by A. S. HOTCHKISS, and manufactured by the SETH THOMAS CLOCK COMPANY, of Thomaston, Conn., are unsurpassed in accuracy of time-keeping and excellence of workmanship. Among the many now running in various parts of the world, we refer to the following:

The Clock in the

City Hall, New York City.
St. George's Church, New York City.
Court House, Chillicothe, Ohio.
U. S. Arsenal, Rock Island, Illinois.
U. S. Arsenal, Indianapolis, Indiana.

Street Clock in front of

T. Kirkpatrick's, 889 Broadway, New York.
Alex. M. Hays & Co.'s, 16th St. and Union Square, New York.
R. Keeping's, 1257 Broadway, New York.
C. H. Phelps', 677 Sixth Avenue, New York.
F. D. Barnum & Co.'s, Memphis, Tenn.
J. H. Colclazer's, Indianapolis, Indiana.
Hallett & Buckland's, Springfield, Mass.
American Clock Co.'s, 581 Broadway, New York.
" " 172 State St., Chicago.
L. Bauman & Co.'s, St. Louis.
Zion's Co-operative Mercantile Inst., Salt Lake City.
B. & O. R. R. Depot, Baltimore, Md.

The Clock in

Reformed Church, Kinderhook, N. Y.
Methodist Church, Derby, Conn.
John Jewett & Sons' Manufactory, Staten Island.
W. Resor & Co.'s Stove Works, Cincinnati, Ohio.
Court House, Mexico, Mo.
Methodist Church, Perth Amboy, N. J.
Onondaga Co. Savings Bank, Syracuse, N. Y.
Naval School Building, Annapolis, Md.
Reformed Church, Saugerties, N. Y.
School House, Ripley, Ohio.
Court House, Meadville, Penn.
Methodist Church, Wilmington, N. C.
E. Jaccard & Co.'s Store, St. Louis, Mo.
Catholic Church, Cold Spring, N. Y.
School House, Lamont, Ill.
Catholic Church, Lancaster, Penn.
Methodist Church, Morristown, N. J.
Odd Fellows' Hall, Portland, Oregon.
Racine College, Racine, Wis.
Presbyterian Church, Babylon, L. I.
Baptist Church, Oil City, Penn.
Lutheran Church, Cumberland, Md.
Universalist Church, Titusville, Pa.
Cornell Watch Co.'s Factory, Chicago, Ill.
Court House, Mansfield, Ohio.
Episcopal Church, Hoosac Falls, N. Y.
Lutheran Church, Johnstown, Penn.
Court House, Freeport, Ills.
Court House, Macomb, Ills.
Baptist Church, Pittsfield, Mass.
Market Building, Houston, Texas.
St. Mary's Church, Brooklyn, N. Y.
Court House, Steubenville, Ohio.
B. C. Wells' Jewelry Store, Austin, Texas.
Mich. So. & R. I. Passenger Depot, Chicago.
————————— Church, York, N. Y.
Asylum, Utica, N. Y.
Knights of Pythias Hall, Gap, Lancaster Co., Pa.
Court House, Greensboro', N. C.
H. W. Collender's manufactory, Stamford, Conn.
Public School House, Nashville, Tenn.
Court House, Freehold, N. J.
Syrian Protestant College, Beirut, Syria.
Germania Savings Bank, Pittsburg, Pa.
School House, Salina, Kansas.
Methodist Church, Long Branch, N. J.
Court House, Sandusky, Ohio.
Dwight Place Church, New Haven, Conn.

And many others

No. 0—TIME—8 DAY.

20 inches wide, 17 inches deep, 42 inches high.

7 inch Main Wheel, of ample power for a Single Dial five feet or less.

4 foot pendulum. Ball weighs 25 lbs.

No. 1—TIME—8 DAY.

25 inches wide, 21 inches deep, 47 inches high.

9 inch Main Wheel, of ample power for Single Dial up to six feet, or four Dials of five feet or less.

4 foot pendulum. Ball weighs 50 lbs.

No. 2—TIME—8 DAY.

30 inches wide, 27 inches deep, 60 inches high.

12 inch Main Wheel, of ample power for single Dial up to ten feet, or for four Dials from five to eight feet.

5 foot pendulum. Ball weighs 50 lbs.

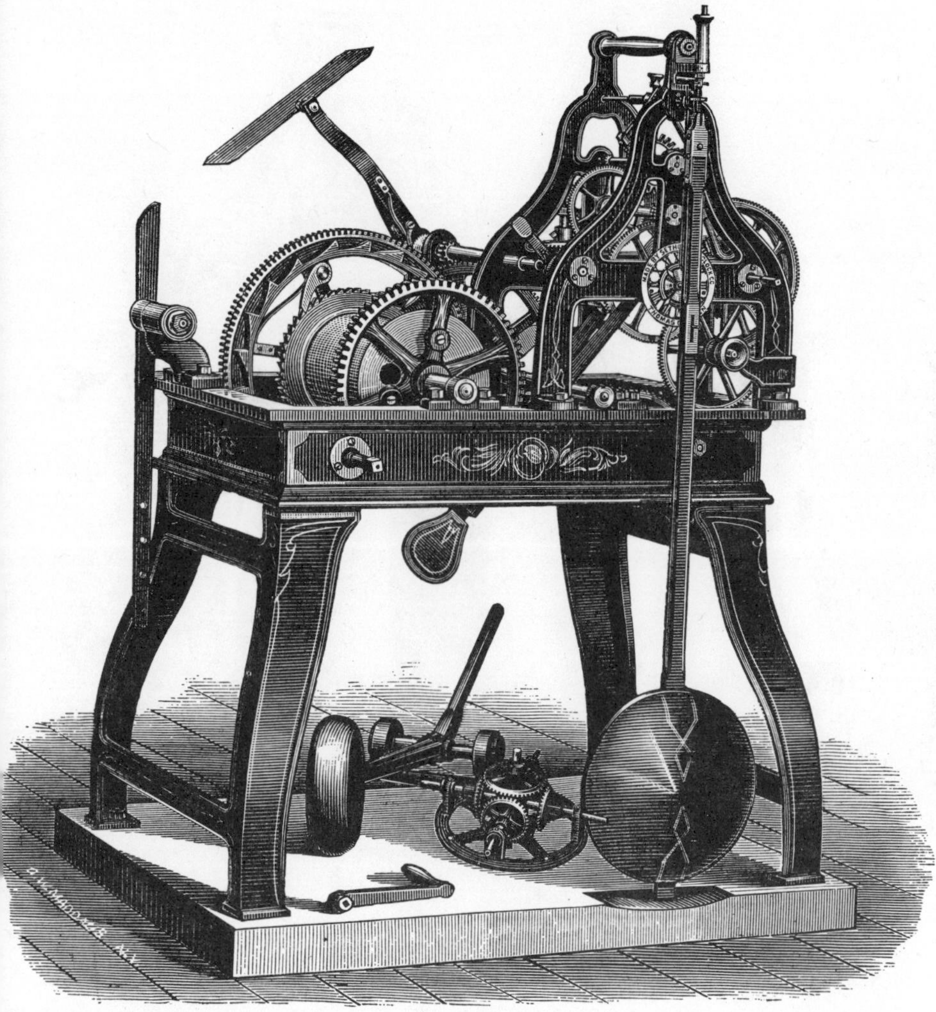

No. 11—STRIKE—8 DAY.

40 inches wide, 42 inches deep, 51 inches high.

9 inch Main Wheel, 16 inch Strike Wheel. For same Dials as No. 1 Time, and striking part of ample power for Bells of 1000 lbs. or less.

5 foot pendulum. Ball weighs 50 lbs.

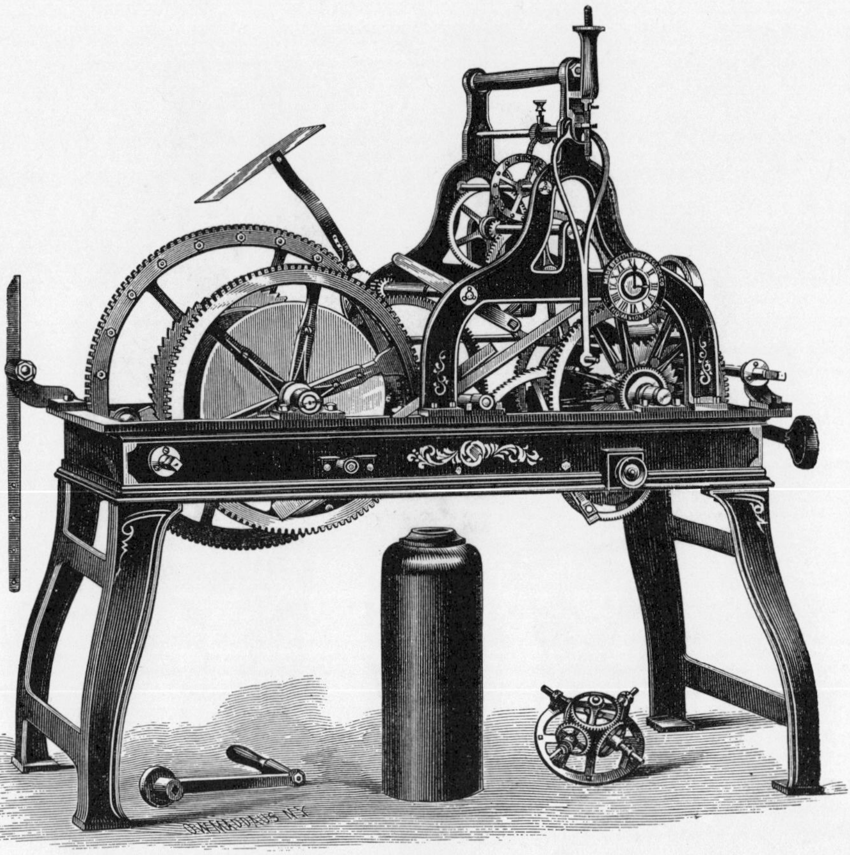

No. 12. A—STRIKE—8 DAY.

48 inches wide, 42 inches deep, 54 inches high.

12 inch Main Wheel, 20 inch Strike Wheel, with Rollers. For any sized Dial, and to strike Bell up to 1500 lbs.

8 foot pendulum. Ball weighs 75 lbs.

No. 12. B—STRIKE—8 DAY.

Same as above, but striking part made heavier and ample for Bell up to 3,000 lbs.

No. 13—STRIKE—8 DAY.

62 inches wide, 50 inches deep, 64 inches high.

16 inch Main Wheel; 24 inch Strike Wheel, with Rollers. For any sized Dials, and for very heavy Bells. All the Wheels made of Bronze, except the main Strike.

8 foot pendulum. Ball weighs 75 lbs.

No. 14—STRIKE—8 DAY.

72 inches wide, 50 inches deep, 70 inches high.

18 inch Main Wheel, 30 inch Strike Wheel, with Rollers. For any sized dial, and for very heavy bells. *All the wheels made of bronze.*

13 foot pendulum. Ball weighs 300 lbs.

TWO-DIAL STREET CLOCK.

Movement to be placed in building, and connection made under sidewalk.

IRON COLUMN.

15 feet to center of Dial. Dial 40 inches.

O.W.MADDAUS N.Y.

FOUR-DIAL STREET CLOCK.

Movement to be placed in building, and connection made under sidewalk.

IRON COLUMN.

15 feet to center of Dial. Dial 36 inches.

WE give below an approximation of the weight required, and the number of feet needed for its fall, to run these Clocks eight days on a single cord:

No. 0.	Time,	25 lbs.	21	feet fall.
No. 1.	"	40 "	30	"
No. 2.	"	65 "	55	"
No. 11.	" side,	40 "	30	"
"	Strike "	175 "	128	"
No. 12.	Time "	65 "	55	"
"	Strike "	175 "	128	"
No. 13.	Time "	80 "	65	"
"	Strike "	200 "	170	"
No. 14.	Time "	125 "	80	"
"	Strike "	250 "	225	"

Where it is not convenient to have so much fall as above, we can arrange to do with one-half, one-third, or one-quarter, by doubling, trebling, or quadrupling the cord on pulleys, and weights must be increased in same proportion.

A. S. HOTCHKISS' TOWER CLOCKS.

These Clocks are made in the best manner, with Jeweled Pallets, Cast-steel Pinions, and Cut Gearing, They can also be furnished to order at special prices to strike the half hours or quarter hours, and also to repeat the hour at the quarters.

Full directions will be sent with each Clock, so that a good mechanic can put them up; or, if desired, an experienced man will be sent to superintend the work, at the expense of the purchaser, unless by special contract.

The within prices include all necessary trimmings for the Clock, (except Dials and Weights), boxed and delivered as directed in New York.

We furnish Hands and Figures (Black or Gilt) adapted to any sized Dial required, but the dials themselves are usually made upon the tower by simply putting on a molding the size required, and painting inside of it.

The Weights can usually be made of wooden boxes filled with stones or bits of iron.

We furnish a Wire Rope for the weights, so constructed that experience has proved it the most durable of anything yet made for the purpose.

In writing for prices and information, please state the size and number of Dials required, the weight of the Bell, and the dimensions of the Tower.

Illuminated Dials of ground Plate Glass, furnished to order, of any size required.

Iron Dials furnished to order, of any size required.

AMERICAN CLOCK COMPANY'S

PRICE LIST

OF

A. S. HOTCHKISS'

TOWER CLOCKS.

SEPTEMBER 1, 1874.

	Arranged for 1 Dial.	Arranged for .... Dials.
No. 0. Time, ...	$ 125 00	$..................
" 1. "	175 00	....
" 2. "	275 00	....
" 11. Strike, ...	350 00	....
" 12. A "	475 00	....
" 12. B "	575 00	....
" 13. "	925 00	....
" 14. "	1,925 00	....

For each additional Dial, add $25 00

Two Dial Street Clock, with No. 1 Time Movement and ordinary Dials, ... $600 00

Two Dial Street Clock, with Glass Dials, for Illuminating, . . 650 00

Four Dial Street Clock, with No. 1 Time Movement and ordinary Dials, ... 800 00

Four Dial Street Clock, with Glass Dials, for Illuminating, . . 900 00

SETH THOMAS
THE CONSERVATIVE, CLOCKMAKING YANKEE

Seth Thomas, son of James and Martha (Barnes) Thomas, was born August 19, 1785, in Wolcott, Connecticut. His father was a cooper by trade and died when Seth was not yet 11 years old.[1] He was apprenticed about 1799 to Daniel Tuttle, a joiner of Plymouth. In 1808, two years after completing his training, he married Tuttle's niece, Philenda (1789-1810).[2]

About 1807 he, along with Silas Hoadley, went to work for Eli Terry who had undertaken the production of 4,000 wood tall clock movements for the Porter brothers of Waterbury. In 1810 Terry sold his interests to Thomas & Hoadley and they continued together manufacturing 30 hour and 8 day wood movement tall case clocks until Thomas sold his interests to Silas Hoadley on December 4, 1813.[3]

Five days later, Seth Thomas bought a clock factory in Plymouth from Heman Clark for $1,500.00 and here he continued to make wood tall clocks under his own name.[4] After Eli Terry developed his shelf clock in 1814, Thomas obtained the right to produce it and did so in box style and pillar and scroll cases.[5] About 1830, Thomas discontinued pillar and scroll production and started producing stencilled column models and later partially and fully carved examples as well as 8 day models. His wood clock production was phased out in 1844 or 1845.[6]

In 1834, Seth Thomas built a cotton sheeting factory, the income from which probably helped him survive the financial panic of 1837 which helped end many other wood clock concerns.[7] By 1860, he was employing 48 people in the mill, but the Civil War made prices so high that the factory was closed.[8]

In 1840, Thomas sent his nephew, Marcus Prince, over to Bristol to learn how Chauncey Jerome was making the cheap 30 hour brass clock.[9] Hiram Camp, Jerome's foreman wrote, "about the year 1840, I think, Seth Thomas, who had not as yet engaged in the making of brass clocks, sent one of his men, a Mr. Prince, over to Bristol to work for Mr. Jerome and learn how to do the work. Mr. Jerome said to his foreman, Mr. Camp, 'Now Mr. Thomas is a good man and he wants to get into making brass movements and I want you to teach Mr. Prince all that you can about the work.' So, after two or three years, Mr. Prince went back to Mr. Thomas, and he began to make the brass movements."[10] Chauncey Jerome himself wrote, "It was never Mr.

Thomas' practice to get up anything new. He never would change his pattern or mode of manufacturing until he was driven to it to keep his customers. At the time when I invented the one day brass clock in 1838, he said much against it, that it was not half so good as a wood clock and that he never would take up anything again that Jerome had adopted, but he was compelled to in a year or two to keep his customers. He sent his foreman to Bristol, where I was then carrying on business, to get patterns of movements and cases... ." [11]

Prince returned to Plymouth Hollow and made the first brass clock movements for Seth Thomas in 1842. In 1844 or 1845 wood movements were phased out and Prince assumed entire control of movement manufacture for Thomas.[12] These earliest brass clocks were probably all in ogee cases with the addition of 30 hour and 8 day cases with columns about 1850. By June of 1850, 24,000 brass clocks, valued at $60,000.00, were being produced annually. Fifty men were employed in the factory and water wheels and a steam engine produced the power.[13]

In 1852, Thomas built a brass rolling mill which produced cast and rolled copper, brass, German silver and other metals.[14] The mill burned a few years later, but was rebuilt before Seth Thomas' death.[15] This business, known as the Seth Thomas Manufacturing Company, was sold May 21, 1869, to Holmes, Booth, & Atwood Mfg. Co. of Waterbury.[16] It survives today as Plume & Atwood Manufacturing Company.

On May 3, 1853, Thomas organized the Seth Thomas Clock Company, a joint stock corporation. He deeded to this company eight pieces of land, one of which included "one dwelling house, two clock factories, engine house, kiln dry barn, and other out buildings ... with the water privelege, steam power engine, waterwheels, fixtures, machines, tools, stock wrought and unwrought and in process." Seth Thomas was the largest of the 33 stockholders and became president.[17]

On Friday night, January 28, 1859, at the age of 73 years, 5 months, and 9 days, Seth Thomas died at his home in Plymouth Hollow.[18] His funeral was held on Sunday and he was buried in the old section of the Hillside Cemetery.[19] His widow, Laura (Andrews) Thomas, who Seth had married as his second wife in 1811, had become "the widow of one of the richest men in the state", according to Jerome. Jerome also wrote a year after Thomas' death, "Seth Thomas was in many respects a first rate man. He never made any improvements in manu-

THREE EARLY BRASS MOVEMENT MODELS, ALL DISCONTINUED BEFORE 1863

FIGURE 1

EARLY FOUR COLUMN, 30 HOUR WEIGHT DRIVEN WITH WOODEN DIAL, WHICH IS UNUSUAL FOR SETH THOMAS. THIS STYLE OF CASE WAS POPULAR ABOUT 1850. (SAMUEL JENNINGS COLL.)

FIGURE 2

EARLY COLUMN MODEL, 30 HOUR WEIGHT DRIVEN. THIS MODEL WAS REPLACED BY SIMILAR ONE WITH OGEE BASE AND TOP MOLDING. SEE PAGES 9-11. (CHRIS H. BAILEY COLL.)

FIGURE 3

EARLY SPRING DRIVEN HALF COLUMN ABOUT 1860. NOTE LARGER TOP AND BASE. THIS MODEL IS 19 INCHES HIGH BUT A 25 INCH, 30 HOUR WEIGHT MODEL WAS ALSO OFFERED BY SETH THOMAS. (ELLIS M. BIDWELL COLL.)

facturing; his great success was in money making. He always minded his own business, was very industrious, persevering, honest, his word was as good as his note, and he always determined to make a good article to please his customers. He had several sons who are said to be smart business men."[20]

Aaron Thomas (1830-1899) took over as president of the company eleven days before his father's death and continued in that capacity until 1892. Aaron was more progressive than his father and introduced many new models in just a very short time. It is believed that at the time of Seth Thomas' death, only 30 hour and 8 day weight clocks were being produced, being in ogee and half-column style cases.[21] Labels on many of these early brass clocks were printed by Elihu Geer of Hartford and data on Geer's activities are shown in Table I to help date these clocks.

About 1860 production began of spring driven clocks in 30 hour and 8 day varieties. Also three styles of Regulators, developed by Silas Burnham Terry, were added about this time.[22] Of these, the Regulator #2, carried with few changes from about 1860 to 1950, is probably the longest produced single model in clock manufacturing history.[23] By 1860, some 40,000 brass clocks were being produced annually.[24]

In 1862, the Seth Thomas Clock Company bought out the patent rights of Huntington & Platts of Ithaca, New York, and added calendar models to their line in late 1862 or in 1863.[25] On February 15, 1876, Seth Thomas' grandnephew, Randall T. Andrews Jr. patented a calendar mechanism that was utilized by the company until the last model, the Parlor calendar #3, was dropped in 1917.[26]

An analysis of Plymouth land records show that Plymouth Hollow had its name changed to Thomaston between March 13, 1865, and June 23, 1865.[27] It was not until July 6, 1875, however, that the town was officially incorporated as Thomaston.[28]

On October 17, 1865, a joint stock company known as Seth Thomas' Sons & Company was formed to produce marine movements and high grade 18 day pendulum movements. This company sold through a New York City outlet and continued until January 31, 1879, when it was consolidated with the Seth Thomas Clock Company.[29]

The exact date of the purchase of A. S. Hotchkiss & Co., manufacturers of tower clocks in New York City, is not known. There is evidence, however, that it was purchased between

November, 1865, and June, 1868.[30] Andrew S. Hotchkiss remained as an employee of the company and set up tower clocks for many years afterwards.[31] By 1900, four distinguished awards had been received for tower clock production. Tower clocks were produced until about 1929, but enough parts were stocked that clocks were installed until the last clock, #3,232, was installed in Lebanon, Connecticut, November 5, 1942.[32]

In 1865, the company entered into the American Clock Company, a sales outlet for Connecticut clocks. Sales outlets included New York City, San Francisco, and Chicago.[33] In 1874, an office had been opened in London.[34] By the year 1870, over 133,000 clocks were being produced annually at the Thomaston factory.[35]

In 1883 a watch shop was constructed and in 1884 the Seth Thomas Clock Company went into the production of pocket watch movements. This continued until 1915.[36] They produced 18 different models, each in 4 grades, utilizing from 7 to 28 jewels.[37]

In 1892, Aaron Thomas retired and his nephew, William Thomas Woodruff, became president. Upon the death of Woodruff in 1919, Seth E. Thomas Jr. (1876-1932), great grandson of Seth Thomas, became president. On January 1, 1931, the Seth Thomas Clock Company became a division of General Time Instruments Corporation, now known as General Time Corporation.[38] The company continues today as a division of that corporation.

Chris H. Bailey
Curator
American Clock & Watch Museum, Inc.

January, 1973

TABLE I

Elihu Geer of Hartford printed most of the labels in Seth Thomas' earliest brass clocks. By 1863 they were being printed by Francis & Loutrel of New York City. Following are the addresses and dates of Geer's activities which can help narrow the age of some of the "Plymouth Hollow" clocks.

1839-1844	26 1/2 State St.
1845-1846	26 State Street
1847-1849	1 State Street
1850-1855	10 State Street
1856-1864	16 State Street
after 1865	18 State Street

REFERENCES

1. HISTORY OF THE TOWN OF WOLCOTT, CONN., 1731–1874, BY REV. SAMUEL ORCUTT. WATERBURY-AMERICAN PRINTING CO., 1874. PP. 352–3, 563.

2. DESCENDANTS OF WILLIAM AND ELIZABETH TUTTLE, BY GEORGE F. TUTTLE. RUTLAND, VT– TUTTLE AND CO., 1883, P. 62.

3. PLYMOUTH LAND RECORDS (PLR), VOL. 4, P. 201.

4. PLR, VOL. 5, P. 25. CLARK WAS PRODUCING 30 HOUR TALL CLOCK MOVEMENTS IN THIS SHOP, NOT HIS 8 DAY BRASS 'SALEM BRIDGE' TYPE.

5. AMERICAN CLOCKMAKING, ITS EARLY HISTORY, BY HENRY TERRY, 1870, P.6.

6. HISTORY OF LITCHFIELD COUNTY, (CONN.) PART 2, PHILADELPHIA – J. W. LEWIS AND CO., 1881, PP. 606–7.

7. HISTORY OF THE TOWN OF PLYMOUTH, BY FRANCIS ATWATER. MERIDAN, CONN.-JOURNAL PUB. CO., 1895, P. 147.

8. OP. CIT. HISTORY OF LITCHFIELD COUNTY, P. 606.

9. IBID., P. 606.

10. A SKETCH OF THE CLOCK MAKING BUSINESS, 1792–1892, BY HIRAM CAMP. WRITTEN MARCH, 1893 (UNPAGED). ORIGINAL COPY IN ACWM LIBRARY.

11. HISTORY OF THE AMERICAN CLOCK BUSINESS, ETC. BY CHAUNCEY JEROME. NEW HAVEN–F.C. DAYTON JR., 1860. P. 103.

12. OP. CIT., HISTORY OF LITCHFIELD COUNTY, PP. 606–7.

13. U.S. FEDERAL INDUSTRIAL CENSUS OF CONNECTICUT, 1850.

14. PLR, VOL. 17, P. 593.

15. OBITUARY OF SETH THOMAS FROM WATERBURY AMERICAN, 4 FEB 1859.

16. PLR, VOL. 27, P. 33.

17. PLR, VOL. 18, PP. 484–5.

18. OP. CIT., OBITUARY.

19. CEMETERY INSCRIPTIONS, HILLSIDE CEMETERY, THOMASTON, CONN.

20. OP. CIT., JEROME, P. 68.

21. SOME 30 HOUR AND 8 DAY SPRING DRIVEN SHELF CLOCKS WERE OFFERED IN THE 1850'S WITH BRASS SPRING MOVEMENTS, PROBABLY PURCHASED FROM THE DEFUNCT FIRM OF TERRY AND ANDREWS OF BRISTOL. ACTUAL SPRING MOVEMENT PRODUCTION PROBABLY DID NOT BEGIN UNTIL AFTER THE DEATH OF SETH THOMAS.

22. OP. CIT., HENRY TERRY HISTORY.

23. SETH THOMAS TRADE CATALOGS IN ACWM LIBRARY.

24. U.S. FEDERAL INDUSTRIAL CENSUS OF CONN., 1860.

25. PATENT MATERIAL IN THE LIFE AND DEATH OF THE ITHACA CALENDAR CLOCK CO., BY JAMES W. GIBBS, SUPPLEMENT TO NAWCC BULLETIN, 1960, PP. 1, 40–55.

26. UNITED STATES CLOCK AND WATCH PATENTS, 1790–1890, BY GEORGE H. ECKHARDT. NEW YORK, 1960. P. 18.

27. PLR, VOL. 24, PP. 46, 49, 66.

28. CONNECTICUT STATE REGISTER AND MANUAL, 1969, P. 579.

29. OP. CIT., HISTORY OF LITCHFIELD COUNTY, P. 603. ALSO PLR, VOL. 22, P. 83, AND 1872 TRADE CATALOG OF S.T.'S SONS AND CO., ACWM LIBRARY.

30. ANALYSIS OF TESTIMONIALS GIVEN IN LATER TOWER CLOCK CATALOGS.

31. NEW YORK CITY DIRECTORIES, 1854-62, 1869, 1873, 1879, CONN. STATE LIBRAR

32. TOWER CLOCK ORDER BOOKS, 1873-1942, AT SETH THOMAS COMPANY.

33. 1865 CATALOG OF AMERICAN CLOCK COMPANY, NEW YORK CITY.

34. LONDON CITY DIRECTORIES SEARCHED BY MR. E.J. TYLER OF NEW MALDEN, KENT, ENGLAND.

35. U.S. FEDERAL INDUSTRIAL CENSUS OF CONNECTICUT, 1870.

36. 'THE SETH THOMAS WATCH SHOP,' BY PAUL G. DARROT, BULLETIN OF NAWCC, VOL. 8, NO. 8, APRIL 1949. PP. 611-13.

37. ALMOST EVERYTHING YOU WANTED TO KNOW ABOUT WATCHES AND DIDN'T KNOW WHO TO ASK, BY GEORGE E. TOWNSHEND, 1970. P. 58.

38. 'A CENTURY AND A QUARTER WITH SETH THOMAS,' PUB. 1938 AS NO. 48 IN A SERIES IN CONNECTICUT INDUSTRY MAGAZINE.

SETH THOMAS WATCH MOVEMENTS

FROM 1904 ILLUSTRATED PRICE LIST